CONTENTS

WHAT ARE FROGS?

Frogs are amphibians, which are animals that usually switch between living in water and living on land during different stages of their life cycle. Frogs have short bodies, no tails and their back legs are typically longer than their front legs, usually with some webbing between their toes. Their eyes tend to be large, sticking out from their heads, and female frogs are almost always larger than the males. They belong to the scientific order 'Anura', which means 'without a tail' in ancient Greek.

The common toad is covered in dry, warty skin and moves with an inelegant walk and short jumps.

What are Toads?

All toads are frogs, but not all frogs are toads. People usually use the term 'toad' for species of frogs that have dry, leathery skin and the ability to live away from water. From now on, when we talk about frogs, this includes toads too.

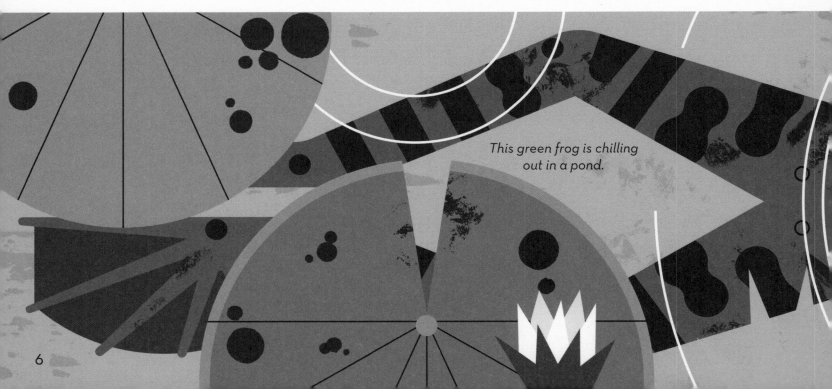

This green frog is chilling out in a pond.

Nom Nom

Adult frogs are usually 'carnivorous', meaning they only eat other animals. They swallow their prey whole, with some species eating any creature that fits inside their massive mouths. A frog's diet primarily consists of insects but can also include worms, snails, mice, fish, snakes, birds, crickets and even small frogs.

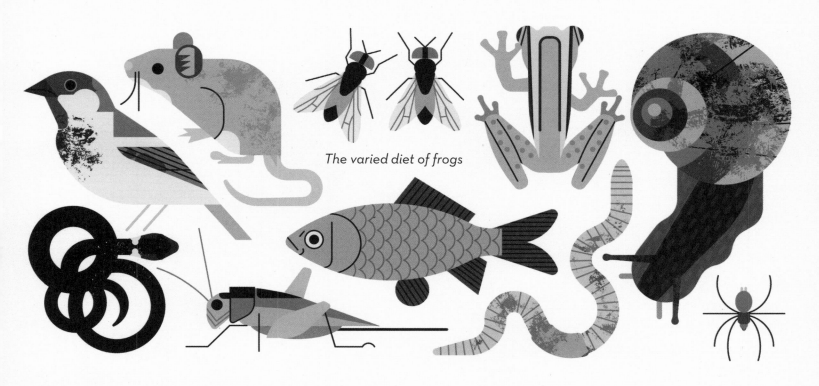

The varied diet of frogs

Here, There and Everywhere

Frogs are found on every continent except Antarctica. They are most abundant in tropical rainforests, but can be found in woodlands, swamps, lakes and even in the Arctic Circle and deserts. Around 7,000 species have been discovered so far, but more are found each year. Many frogs are small, well camouflaged and can hide in difficult-to-reach areas, so it's not surprising we haven't found them all yet!

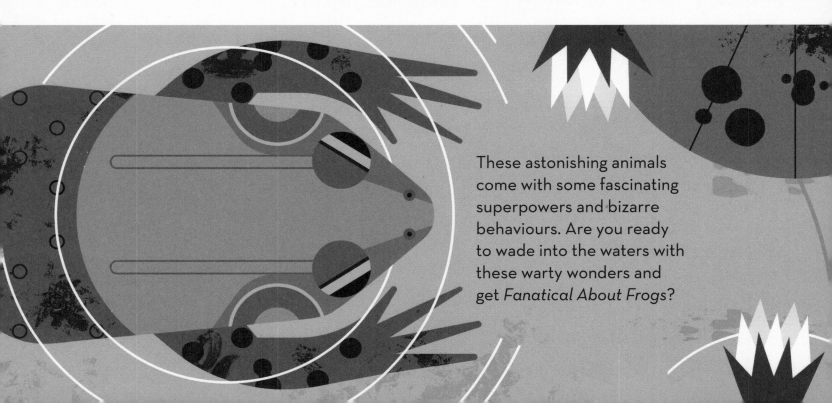

These astonishing animals come with some fascinating superpowers and bizarre behaviours. Are you ready to wade into the waters with these warty wonders and get *Fanatical About Frogs*?

WARTS AND ALL

This northern leopard frog showcases many of the special features and abilities for which Anurans are known.

Tympanum

Frogs do not have external ears like us. Many frogs have a circular piece of skin called a tympanum instead. This covers their eardrums, making them much more waterproof.

Nictitating Membrane

This thin inner eyelid keeps the eye moist, protects it from dust and helps the frog see clearly underwater.

Eyes

Sight is very important to frogs. Their eyes don't rotate like ours do, but stick out from their head. This gives them a very wide field of vision – almost 360 degrees! The shape of their pupils varies depending on the species.

Horizontal Oval

Colorado River toad

Vertical Oval

Red-eyed tree frog

Heart-shaped

European fire-bellied toad

Diamond

Purcell's ghost frog

Circular

Madagascar tomato frog

Warts

Some species of frog, especially those commonly called 'toads', are covered in large bumps. Smaller bumps produce poisonous or bad-tasting chemicals, while larger, balloon-like sacs near the head contain a concentrated dose of poison. Many people believe that touching frogs can give you warts, but this is not true.

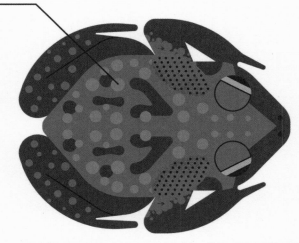

The poison contained in the skin of this warty cane toad can cause illness, hallucination or even death.

Feet

Many frogs have webbing between their toes, so their feet act like flippers while swimming. Other frogs, such as those who live in dry areas or in trees, usually have only partially webbed toes or no webbing at all. However, some tree frogs have excessively webbed toes, which they use as parachutes to glide between trees.

Wallace's flying frogs can glide up to 15 metres between trees.

Adhesive Pads

Some frogs have enlarged sticky pads at the tips of their fingers and toes to help them climb trees and hang out upside down.

African dwarf frogs have webbing between their toes because they spend most of their lives underwater.

Legs

Many species of frog can jump very far using their long, powerful hind legs. An elastic tendon wraps around the frog's ankle bone and pings off to propel the frog forwards, like an arrow from a bow.

Tubercle

Species that burrow into the ground have hardened skin nodules sticking out from the heels of their hind feet.

The black tubercle on this Eastern spadefoot toad's foot helps it burrow underground, where it spends most of its life.

SKIN DEEP

Frogs have 'semipermeable' skin, which means it can absorb water and gas. Most frogs breathe like humans, but their skin also allows them to absorb oxygen from water. This enables them to stay submerged for long periods without having to come up for air. They also don't need to drink to stay hydrated, as their body does it for them. The downside is that frogs can lose water through their skin, making them vulnerable to drying out. This is why most frogs live near water, as this keeps them moist.

The Bornean flat-headed frog has no lungs and breathes entirely through its skin.

Jump Out of Your Skin

Did you know that frogs shed their skin? Some do it daily, while others do it weekly. The outer layer of skin splits down the belly and back, then the frog pulls it off with its legs. Most frogs then eat the old skin, as it contains protein and other nutrients.

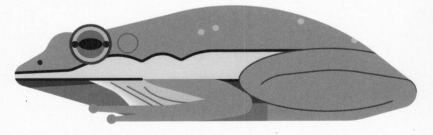

This American green tree frog sheds its skin and eats it.

Snot On

Many frogs are slippery. Their skin produces mucus, similar to the stuff that comes out of our noses, and this helps keep their skin wet. This mucus can sometimes contain chemicals too, which protect frogs against bacteria, fungi and, in some cases, from being eaten by predators.

If you picked up a yellow dyer rainfrog, it would feel slimy and your hands would be dyed yellow!

A Place in the Sun

Some frogs apply their own sun cream during dry, sunny periods. They produce a waxy substance from glands in their necks, then rub this all over their bodies using their legs. The wax holds in moisture and protects the frog from drying out.

This waxy monkey tree frog makes sure its wax covers even the hard-to-reach places.

A Little Extra

Some frogs have 'dorsolateral folds' along their backs. These ridge-like folds are a special adaptation to give the frogs' skin more surface area, because more skin means more oxygen from water.

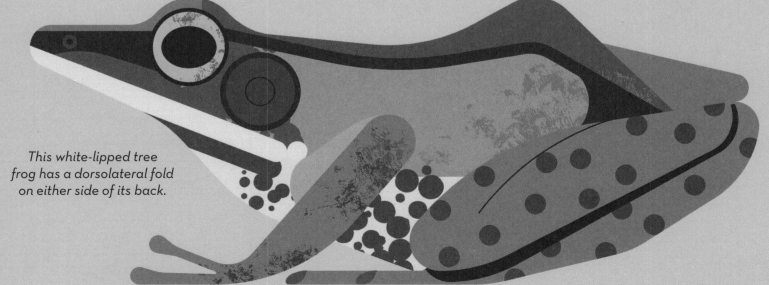

This white-lipped tree frog has a dorsolateral fold on either side of its back.

MAKING A MEAL OF THINGS

Many frogs are superbly accurate when targeting prey. This is thanks to a secret weapon: their tongue. It shoots out, sticks to the insect, or other unfortunate creature, and then drags it back into the frog's mouth. This movement is so fast that it can happen within the blink of an eye.

Roll Off the Tongue

A frog's tongue is usually attached at the front lower part of its mouth and is coiled back towards the throat. When ready, the frog opens its mouth and the tongue unfolds so speedily that it is thrust at the prey like a spring.

Big Mouth

Frogs have exceptionally wide mouths, which allow them to swallow their prey whole. Many have sticky tongues to catch unsuspecting victims, but sometimes they simply stuff creatures into their mouths using their hands.

A green frog catches a dragonfly

On the Tip of Your Tongue

A frog's tongue is soft and covered in very special spit. When still, the spit is super sticky and firm, like honey. But while in motion, the force of the quick tongue action turns it into a runny liquid. When a frog's tongue hits its target, the saliva spreads over the surface of the prey. As it stops moving, it very quickly firms up again and traps the victim on the end of the tongue.

Close Your Eyes

Have you ever seen a frog swallow? Curiously, it closes its eyes to do so. What actually happens is that the frog's eyeballs are pulled into its mouth briefly to push the prey against its tongue. This loosens the spit again, allowing the frog to swallow its meal more easily.

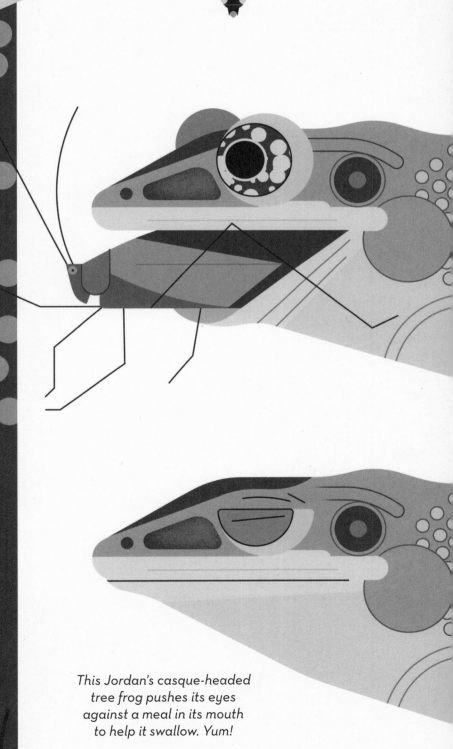

This Jordan's casque-headed tree frog pushes its eyes against a meal in its mouth to help it swallow. Yum!

DRESS FOR SUCCESS

Frogs come in a stunning variety of colours and patterns, sometimes even within a single species. But these aren't just fashion statements. Many skin pigments and patterns have a specific purpose to help our froggy friends survive.

Hide and Seek

'Camouflage' allows an animal to blend in with its surroundings, making it hard to see even if it's right in front of you. Lots of frogs have camouflaged skin, which keeps them hidden from predators that hunt using eyesight, like birds and fish. Some frogs even pretend to be something else entirely.

This stripeless tree frog is the same colour as the leaf it's sitting on.

Projections above the nose and eyes of these long-nosed horned frogs make them look like leaves.

The aptly-named Vietnamese mossy frog hides in plain sight against moss.

Pied warty frogs sit still and pretend to be bird poo

Dangerous Fashion

Certain frog species have bright, colourful and dramatic contrasting patterns on their skin, which convey a very simple message: 'I am not safe to eat'. Usually, these species have skin secretions that may be toxic to predators or simply make them taste terrible. Creatures that use colours as a warning are called 'aposematic'.

14

What's Your Poison?

The most famous family of aposematic frogs are the poison dart frogs of Central and South America. The skin secretions of these little amphibians contain some of the most poisonous substances on the planet. Some indigenous tribes in these regions use the poison to coat the tips of their blowgun darts.

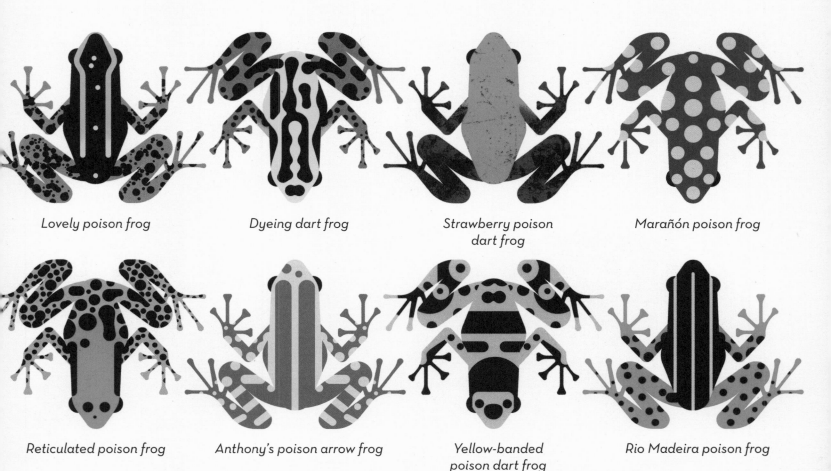

Lovely poison frog　　*Dyeing dart frog*　　*Strawberry poison dart frog*　　*Marañón poison frog*

Reticulated poison frog　　*Anthony's poison arrow frog*　　*Yellow-banded poison dart frog*　　*Rio Madeira poison frog*

Best of Both Worlds

The eastern grey tree frog can change its skin from green to grey to brown to blend in with different surroundings. However, when it jumps, a bright orange 'flash mark' is revealed on its legs, which is thought to startle or confuse predators just long enough to allow the frog to escape.

This grey tree frog's skin is brown when it sits on a lichen-covered ash tree.

The same grey tree frog can change its skin to be green to blend in with a leaf.

The underside of the grey tree frog shows the bright orange patches on its legs.

HOT AND COLD

Frogs are 'ectothermic'. This means they can't control their internal body temperature like humans do. To regulate their temperature, they have to change their behaviour. They may rest on something warm or bask in the sun to heat up, or hide in the shade and sit in water to cool down. Some frogs regulate their temperature by darkening or lightening their skin. Pale colours reflect light and heat away from the body, while darker colours absorb them.

Night Life

Most frogs are 'nocturnal', meaning they are much more active when the sun goes down. During the day, these frogs rest in a safe, cool place until nightfall.

Starry night reed frogs are white during the day and black at night.

This Australian green tree frog is chilling out in a tree hole, waiting for the night shift.

The southern leopard frog typically hibernates underwater.

Hide and Sleep

When winter rolls around, frogs move less to conserve their energy. Some species will find a place to retreat, hidden from predators and harsh weather, then slow down the processes in their bodies that require energy. This state of 'dormancy' lasts until the weather improves in spring.

Croaked It?

Some frogs hang out in logs or under rocks during winter, with little protection from the cold. When the temperature drops, the frogs begin to freeze: they stop breathing and their hearts stop beating. They appear dead, but amazingly, these frogs survive thanks to a natural antifreeze in their bodies. After temperature increases again, the frogs thaw out and get on with their lives as if nothing happened.

A frozen wood frog

A Main's frog aestivating in its skin cocoon

Shed and Buried

Some frogs go into a dormant state known as 'aestivation' during hot or dry spells. Burrowing into the soil, these frogs shed several layers of skin to create a nearly waterproof cocoon around their bodies. When it gets more bearable above ground, the frogs resurface for a little while. Some desert frogs live their entire lives this way, only emerging when they sense rain.

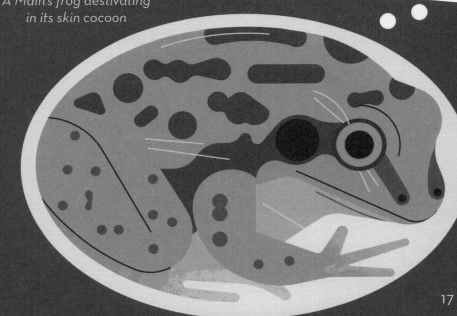

GIVE ME A CALL

To woo females or ward off rivals, male frogs produce calls that are unique to each species. Sometimes these vocalisations are the only way to tell the difference between two species that look physically identical. Males usually make these calls near water or within territories that provide a suitable place for a female to lay her eggs.

Carrying a Tune

Many male frogs have one or more vocal sacs, which expand to amplify their 'music'. Frogs croak, bark, purr, moo, chuck, bleat, quack and click. Their calls can sometimes be heard over a kilometre away.

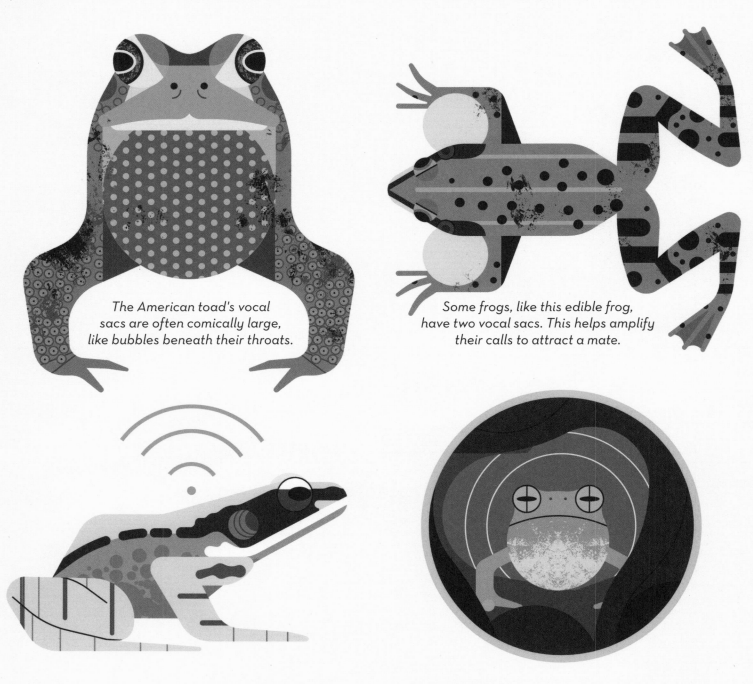

The American toad's vocal sacs are often comically large, like bubbles beneath their throats.

Some frogs, like this edible frog, have two vocal sacs. This helps amplify their calls to attract a mate.

Concave-eared torrent frogs deposit their eggs by loud rushing water. These sounds drown out normal calls, so they use ultrasonic communication instead. Their call is so high-pitched, humans can't hear it.

The Borneo tree-hole frog sits inside tree holes to amplify its calls. Based on the size of the hole and the water inside, the frog adjusts the sound's pitch for maximum efficiency.

Different Croaks for Different Folks

Some frogs make different noises depending on their audience. They may use one call to serenade a warty beauty and a different one to tell the other guys to back off. Another sound, known as the 'fright scream', may be used as a distress call when they feel endangered. While frogs produce most calls with their mouths closed, they emit fright screams with their mouths open.

American bullfrogs sound like a cow when trying to attract a mate...

...but when threatened, they let out a scream similar to a human's.

Lights, Camera, Ribbit

If frogs have so many different calls, why do we think that frogs 'ribbit'? The simple answer is because of Hollywood. Movies have been produced there since the early 1900s and scenes set at night would often be filmed in the local area, where the calls of frogs could be heard in the background. The Pacific tree frog and the California tree frog are common in the area and both make this classic 'ribbit' noise.

Two secret stars – the Pacific tree frog (right)...

...and the California tree frog (left)

BORN THIS WAY

Frogs go through several massive changes as they grow up. At each stage of their life cycle, frogs will look and behave differently. The process of changing from egg to adult is called 'metamorphosis', and it happens in a variety of remarkable ways across different species.

This life cycle shows the development of the European common frog.

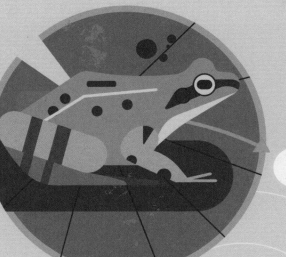
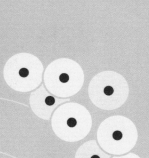

Adult

When a froglet begins to leave the water and absorbs the remains of its tadpole tail into its body, it becomes a frog. After a couple of years, this frog will be able to mate and produce offspring of its own.

Egg

Frogs lay soft, jelly-like eggs, which are stuck together in a group called a 'clutch'. The eggs are vulnerable to drying out, so they are laid in watery areas, such as ponds. A single clutch can contain thousands of eggs.

Froglet

The tadpole grows back legs first, and then a few weeks later its front legs appear. The tail begins to disappear too, as it develops into a tiny version of an adult frog. It has now changed into a froglet.

Tadpole

A tadpole wriggles free from its egg and begins swimming around the water. Tadpoles are similar to fish, with strong tails for swimming and gills in their heads, which help them breathe underwater.

The common toad lays its eggs in a string formation.

Some frogs skip the free-roaming tadpole stage in their development. This allows them to breed on land. In this 'direct development' cycle, the transformation happens within the egg and a fully formed froglet emerges once it is ready.

These two Fiji tree frog eggs don't hatch as tadpoles...

...tiny froglets are ready to come out instead.

Doing Things Differently

Beating your eggs may sound like bad parenting, but grey foam-nest tree frogs come together in groups of up to 30 to beat their eggs together with fluid using their feet. This forms a foam, which hardens, protecting the developing eggs.

When Darwin's frog tadpoles emerge, the male scoops them up in his mouth. The tadpoles then complete most of their growth cycle inside his vocal sac and six weeks later, he throws up lots of little froglets.

The female Suriname toad releases eggs that are then embedded in her back by her partner. Skin forms over them to create little pockets where the eggs develop. Eventually, miniature froglets will burst out of her back. Yuck!

Strawberry poison dart frogs lay their eggs on leaves. The male guards them, keeping them moist, and when the tadpoles emerge, the female carries them one by one to different water sources. She then returns to each one every few days to bring them some food.

IT'S A FROG'S LIFE

Featured Creatures: Red-Eyed Tree Frog

These iconic frogs are easily recognisable with their vibrant red eyes, bright green backs, yellow-blue markings and orange feet. As their name suggests, they spend most of their lives in trees. This is called an 'arboreal' lifestyle. They are superb at jumping, and their mucus-covered toe pads allow them to stick to most surfaces, including branches, leaves and even glass.

This colourful chap is a red-eyed tree frog.

Red-eyed tree frogs are nocturnal, so they are more active at night. During the day, they find a safe spot on a leaf and go into camouflage mode. They pull their legs in tight to hide their vivid markings, tuck their feet under their body and close the nictitating membrane over their eyes. This allows them to see without exposing their brightly coloured eyes, which would give away their location to potential predators.

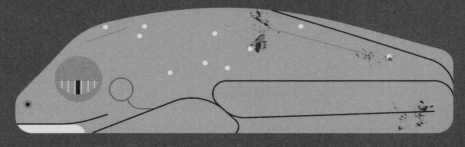

Red-eyed tree frogs appear almost fully green when in camouflage mode...

If the frog is spotted – it quickly opens its eyes and jumps away. The bright yellow and blue parts of the frog's body are said to mesmerise the predator momentarily, causing it to lose sight of its prey, giving the frog time to escape.

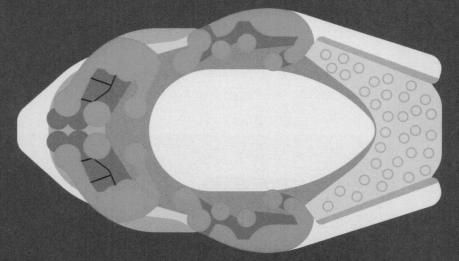

...but underneath they're hiding something far more colourful.

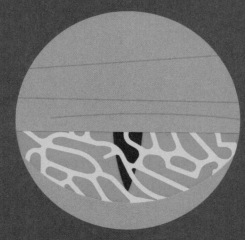

Check out this red-eyed tree frog's eye, covered with its nictitating membrane.

Stuck Like Glue

Red-eyed tree frogs produce a sticky jelly to glue their eggs together and attach them to a leaf above a puddle or pond. It takes about a week for the tadpoles to develop and fall into the water below. If attacked by a predator though, a tadpole may emerge early to increase its chance of survival. When this happens, the vibrations of the tadpole's rapid tail movement signals danger to its siblings, and the whole clutch begins to plunge into the water.

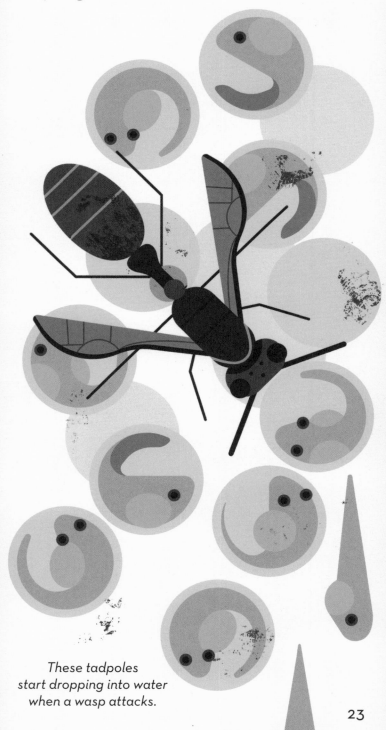

These tadpoles start dropping into water when a wasp attacks.

Shake it Off

Male red-eyed tree frogs shake their butts to win an argument. They claim territory by perching on a branch and shaking it violently. This produces vibrations that travel through the plant saying, 'This is mine!'. If another male turns up, they'll both shake their branches until one gives in or they start a mini wrestling match.

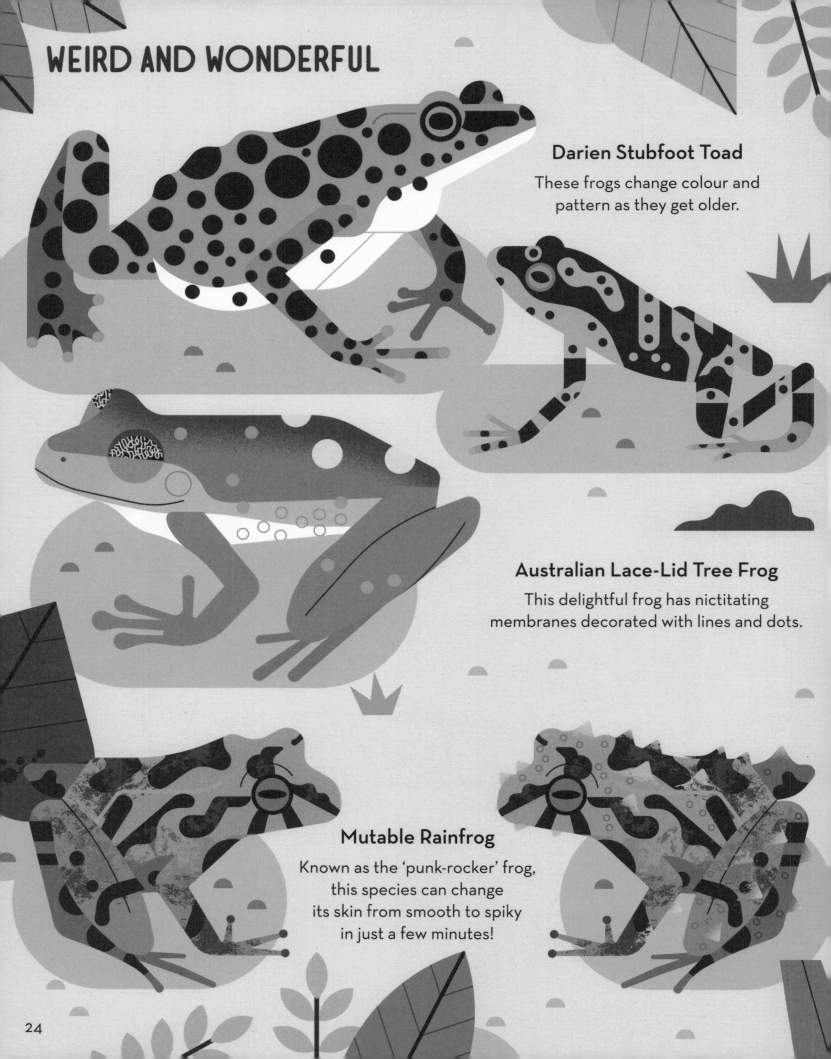

WEIRD AND WONDERFUL

Darien Stubfoot Toad

These frogs change colour and pattern as they get older.

Australian Lace-Lid Tree Frog

This delightful frog has nictitating membranes decorated with lines and dots.

Mutable Rainfrog

Known as the 'punk-rocker' frog, this species can change its skin from smooth to spiky in just a few minutes!

Desert Rain Frog

The adorable angry squeak of this cute frog is worth looking up – there's nothing quite like it!

Purple Frog

This rather strange-looking creature spends most of its life underground.

Brazilian Poison Frog

This stunningly beautiful species of poison dart frog looks like it's wearing fancy pyjamas.

Botta's Bright-Eyed Frog

These frogs are unique to the rainforests of Madagascar and have the most spectacularly colourful eyes.

25

LITTLE AND LARGE

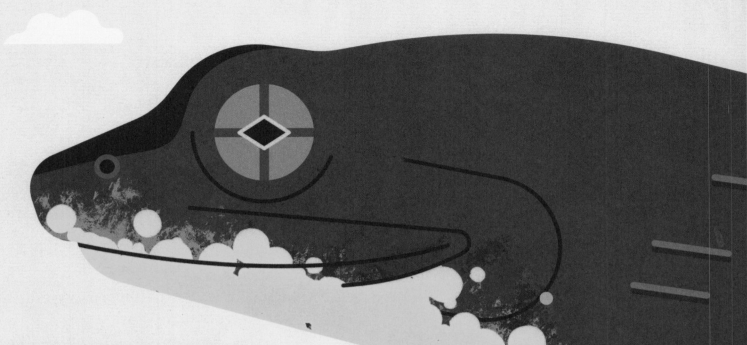

Featured Creatures: Goliath Frog

Goliath frogs are the largest species of frog in the world. Found in central Africa, these beasts can reach 32 centimetres from nose to rear and can weigh more than 3 kilograms – around the same as a newborn human baby! Males tend to be larger than females, which is rare in frogs. Despite their huge size as adults, Goliath frogs start life as modest-sized tadpoles, reaching only around 5 centimetres in length. Adults may feed on anything from dragonflies and worms to crabs, snakes and baby turtles.

*A tadpole
of the goliath frog*

*A life-size illustration
of a goliath frog*

Featured Creatures: Amau Child Frog

These frogs only grow to less than a centimetre long and are the world's smallest vertebrate (an animal with a backbone inside its body). With a species this small, it's not surprising that scientists in Papua New Guinea didn't discover it until 2010!

A life-size illustration of an Amau child frog

TO SCALE

Frogs come in all shapes and sizes. The frogs on these pages are shown at their common life-size.

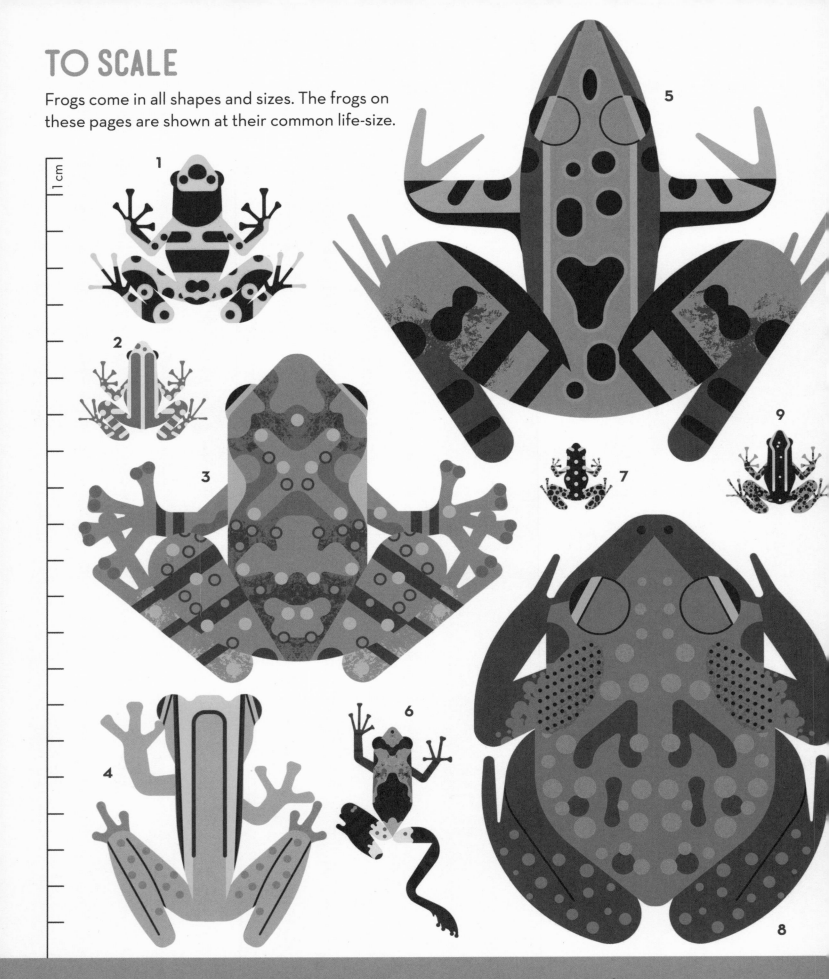

1cm

1 *Yellow-banded poison dart frog* 2 *Anthony's poison arrow frog*
3 *Vietnamese mossy frog* 4 *Porto Alegre golden-eyed tree frog* 5 *Northern leopard frog*
6 *Pied warty frog* 7 *Brazilian poison frog* 8 *Cane toad* 9 *Lovely poison frog*

10 *Long-nosed horned frog* **11** *Rio Madeira poison frog* **12** *Dyeing dart frog*
13 *Marañón poison frog* **14** *Edible frog* **15** *Wallace's flying frog* **16** *Strawberry poison dart frog*
17 *Starry night reed frog* **18** *Stripeless tree frog* **19** *Reticulated poison frog*

AND THE AWARD GOES TO...

Glass frogs have very thin skin on the underside of their torso, which allows us to see inside their bodies! One species, the Yaku glass frog, wins the award for the most transparent frog. You can see its muscles, veins, skeleton, and organs through its skin. You can even watch its heart beat or its stomach digest food.

The award for the loudest frog goes to the Puerto Rican common coquí frog. Males call through the night with a noise as loud as a lawn mower. Their call has two distinct tones that give the frog its name – 'co' and 'kee'. The first tone drives away rival males and the second tone is for enticing females.

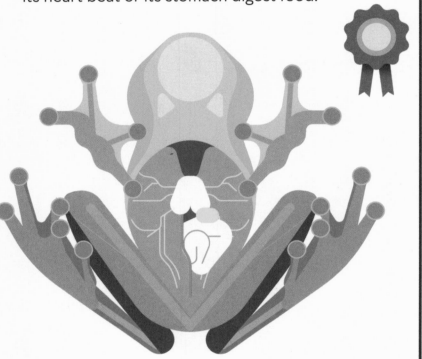

The mimic poison frog wins the award for best impersonator. This harmless species mimics the colouring of three different poisonous frogs so predators will avoid it.

The superbly weird wolverine frog has to win the creepiest frog award. It has 'hairy' sideburns and, to fight off threats, it intentionally breaks its toe bones so they burst through its skin like makeshift claws. The 'hairs' on males are actually little bits of skin, which are thought to increase its surface area and help it absorb more oxygen.

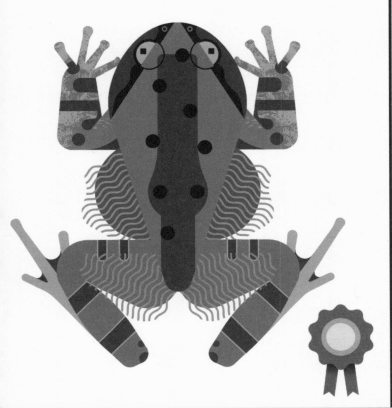

Many poison arrow frogs produce dangerous toxins, but none are more deadly than the golden poison arrow frog. The stuff that oozes from their skin is 20 times more lethal than that of any other frog. Just one gram of this poison is enough to kill several thousand people, so it's not surprising that scientists wear thick gloves when picking them up.

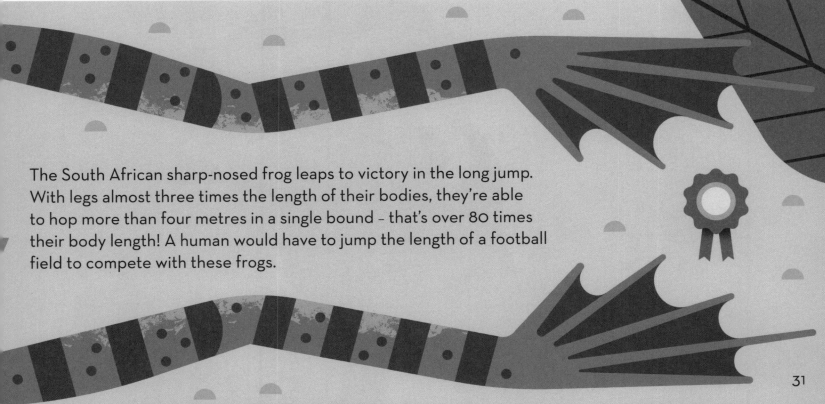

The South African sharp-nosed frog leaps to victory in the long jump. With legs almost three times the length of their bodies, they're able to hop more than four metres in a single bound – that's over 80 times their body length! A human would have to jump the length of a football field to compete with these frogs.

FROG MYTHOLOGY

Jin Chan

In this ancient Chinese myth, the 'Money Frog' is said to appear during a full moon to signify the coming of good news. In works of art it is usually shown as a three-legged bullfrog with red eyes and a coin in its mouth, sitting on a pile of money.

Jin Chan

A Panamanian golden frog

Panamanian golden frog

Although thought to be extinct in the wild, Panama legend has it that the Panamanian golden frog turns to gold when it dies and anybody who spots it will have good fortune. Its image appears on everything from architecture to lottery tickets and indigenous people in the region decorate clothes with the frog's likeness. August 14th is 'National Golden Frog Day' in Panama.

Jiraiya

In Japanese folklore, Jiraiya was a ninja who could control water and command frogs. He also had the power to turn himself into a gigantic toad, but is often shown riding around on one instead.

Jiraiya riding his toad

The Frog Prince

This fairy tale is about a frog who magically transforms into a handsome prince after meeting a pampered princess. Nowadays, the story goes that the frog prince changes after the princess kisses him, but in early versions it's said she throws him against a wall instead!

Around a third of known frog species are under threat and over 100 are now considered extinct. Frog populations are plummeting as a result of pollution, habitat loss, invasive species and even a deadly fungus. Frogs are a very important part of the ecosystems they live in, both as a predator and as a food source, so it's extremely important that we protect our world's frogs. They are eaten by everything from birds, snakes and fish to flesh-eating plants and, of course, bigger frogs.

To Eat or Be Eaten

The Túngara adult frog and tadpole lead separate lives, each with different diets and predators. This food web is an example of how this frog fits into its ecosystem, but all frogs are different and little is actually known about exactly what they eat and what eats them.

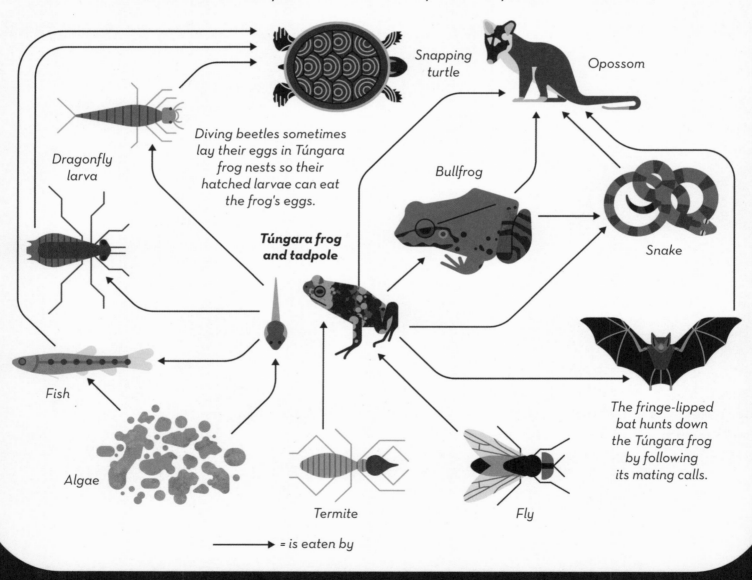

Snapping turtle

Opossom

Dragonfly larva

Diving beetles sometimes lay their eggs in Túngara frog nests so their hatched larvae can eat the frog's eggs.

Bullfrog

Túngara frog and tadpole

Snake

Fish

Algae

Termite

Fly

The fringe-lipped bat hunts down the Túngara frog by following its mating calls.

→ = is eaten by

One of the largest problems facing frogs globally right now is the spread of 'chytridiomycosis' – a fungal disease that has been known to infect and wipe out entire frog populations. Some species have bacteria on their skin that makes them resistant to the disease, so biologists are trying to figure out

A pet Argentine horned frog in its tank

What Can We Do?

If you and your family are interested in keeping a frog as a pet, make sure you look for a reputable dealer who breeds their own. Some pet stores import wild-caught frogs, which may be infected with chytridiomycosis. Buying frogs that were not raised in captivity harms their already declining populations, too – especially since most of the frogs caught in the wild don't survive the journey to the store.

Troubled Waters

Toxic pesticides are often used on gardens and crops to help protect them from insects. These pesticides can be washed into waterways, where frogs are likely to absorb these harmful pesticides through their skin. You can help prevent this by growing your own crops or by buying organic food, which is grown without the use of harmful chemicals.

Organic fruit and vegetables

Leave Well Alone

If you see a frog anywhere, feel free to look at it and study what it does, but make sure you leave it where it is. Please don't pick it up, especially if you have sun cream or insect repellent on your hands – those will easily absorb through a frog's skin and make it ill.

INDEX

CR = *critically endangered*   NT = *near threatened*
EN = *endangered*   VU = *vulnerable*

If you like this, you'll love...

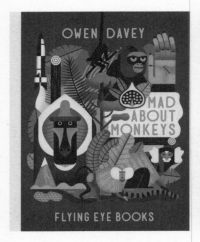

ISBN 978-1-909263-57-4

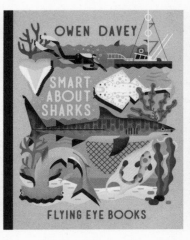

ISBN 978-1-909263-91-8

ISBN 978-1-911171-16-4

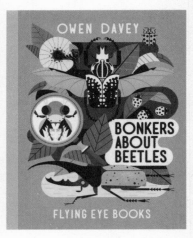

ISBN 978-1-911171-48-5

For my little froglet, Lyra.

Fanatical About Frogs is © Flying Eye Books 2019.

First edition published in 2019 by Flying Eye Books,
an imprint of Nobrow Ltd. 27 Westgate Street, London, E8 3RL.

Text and illustrations © Owen Davey 2019.

Scientific consultant: Rebecca Marie Brunner

Every attempt has been made to ensure any statements written as fact have been checked
to the best of our abilities. However, we are still human, thankfully, and occasionally
little mistakes may crop up. Should you spot any errors, please email info@nobrow.net.

1 3 5 7 9 10 8 6 4 2

Published in the US by Nobrow (US) Inc.

Printed in Latvia on FSC® certified paper.

MIX
Paper from
responsible sources
FSC® C002795

FSC
www.fsc.org

ISBN: 978-1-912497-05-8

www.flyingeyebooks.com